AF428054

Improved Chicken Interventions and Women Empowerment in Bariadi and Muheza Districts, Tanzania

Regina M. Maunde

Title: Improved Chicken Interventions and Women Empowerment in Bariadi and Muheza Districts, Tanzania

ISBN: 979-8-89248-129-8

Author: Regina M. Maunde

Cover image: https://pixabay.com/

Publisher: Generis Publishing
Online orders: www.generis-publishing.com
Contact email: info@generis-publishing.com

ABSTRACT

Women empowerment has recently become a global policy objective, thus attracting much attention. The Sustainable Development Goals (SDGs) recognize women empowerment as one of the 2030 agenda. In the same vein, development partners across the world, including governments, Non-Governmental Organizations (NGOs) and International Development Organizations (INGOs), are increasingly concerned with women empowerment. Development policy makers and practitioners have realized the importance of empowering women as a way to enhance agricultural production and alleviate rural poverty. As such, more and more development projects are incorporating this issue and integrating activities designed to empower women into their structures and the implementation processes. However, evidence on whether specific development projects really empower women is limited. Various studies have revealed that projects that target women succeed to increase their access to income, but do not necessarily raise their empowerment levels. Therefore, this study evaluated the role of improved chicken intervention in empowering women socio-economically in Bariadi and Muheza districts, Tanzania. The study specifically: (i) analysed the determinants of women empowerment in relation to improved chicken intervention, (ii) determined the contribution of the African Chicken Genetic Gain (ACGG) project to women empowerment, and (iii) assessed the role of Community Innovation Platforms (CIPs) in promoting women empowerment. The study adopted a cross-sectional research design and involved 240 women as its respondents. Both qualitative and quantitative data were collected. The qualitative data were collected through Focus Group Discussions (FGDs) and Key Informant Interviews (KIIs) while the quantitative data were collected using a structured questionnaire. Content analysis procedures were used to analyse the qualitative data. The quantitative data were analysed using descriptive statistical analysis, index scales, Composite Empowerment Index (CEI) and ordinal logistic regression. The findings indicated that the majority of women in the two districts had achieved only a medium level of empowerment. The ACGG beneficiaries were more empowered (CEI = 0.714) than non-beneficiaries (CEI = 0.529). Results from ordinal logistic regression showed that age, the form of marriage, religion and involvement in the ACGG project significantly influenced women empowerment ($p < 0.05$). According to the findings, the ACGG project gave women access to income and an opportunity to challenge the cultural norms and practices that influence power relations. Keeping of improved chicken breeds was found to be a new innovation in the study area. Meanwhile, Community Innovation Platforms (CIPs) provided opportunities for collaboration not only to the stakeholders along the chicken value chain, but also to the potential stakeholders responsible for gender integration and community development at large. The study recommends that the Local Government Authority (LGA), Non-Governmental Organizations (NGOs) and private sectors through their development projects should implement similar projects and extend them to other areas. Also, they should design educational programmes intended to get rid of harmful cultural norms and religious beliefs which impede women empowerment. Moreover, development partners should employ the Innovation Platforms (IPs) approach in their implementation of intervention projects, particularly those aimed at empowering women.

TABLE OF CONTENTS

LIST OF TABLES

LIST OF FIGURES

LIST OF ABBREVIATIONS AND ACRONYMS

ACGG	African Chicken Genetic Gain
AGREST	Agricultural Economics Society of Tanzania
AIS	Agricultural Innovation Systems
AR4D	Agricultural Research for Development
CARE	Cooperative for Assistance and Relief Everywhere
CBO	Community Based Organization
CEI	Composite Empowerment Index
CHANGE	Creating Homestead Agriculture for Nutrition and Gender Equity Project
CHIF	Community Health Insurance Fund
CIPs	Community Innovation Platforms
Co-PI	Co-Principal Investigator
DCDO	District Community Development Officer
DCI	Domestic Consultation Index
DLFOs	District Livestock and Fisheries Officers
ECLAC	Economic Commission for Latin America and the Caribbean
EDM	Economic Decision Making
FGDs	Focus Group Discussions
FMI	Freedom of Movement Index
HDMI	Household Decision Making Index
HKI	Helen Keller International

INGOs	International Non-Governmental Organization
IP	Innovation Platform
LGA	Local Government Authority
NGO	Non-Governmental Organization
NSGD	National Strategy for Gender Development
NSGRP	National Strategy for Growth and Reduction of Poverty
OECD	Organization for Economic Cooperation and Development
PAI	Personal Autonomy Index
PO	Postal Office
SDG	Sustainable Development Goals
SPSS	Statistical Package for Social Sciences
TDV	Tanzania Development Vision
TSAP	Tanzania Society of Animal Production
UN	United Nations
UNDP	United Nations Development Programme
UNECE	United Nations Economic Commission for Europe
URT	United Republic of Tanzania
USAID	United States Agency for International Development
VC	Value Chain
WEI	Women Empowerment Index
WIII-FM	What Is In It For Me

1.0   INTRODUCTION

1.1   Background information

Gender inequality is one of the critical challenges that constrain the efforts towards attainment of Sustainable Development Goals (SDGs) (UN, 2015). It is a widespread issue across all communities in the world (Bayeh, 2016). This problem reported to be more pronounced in most of the developing countries, including Tanzania, where women are among the most disadvantaged (Fox, 2018), especially in terms of power balance in different domains. Power imbalance in those nations is high partly due to the widespread gender-based inequalities sanctioned by the existing customs, traditions and laws (van Eerdewijk, 2017). Power imbalance or disempowerment is evidenced by the lower status of women in various domains, especially in education, ownership and control of income, bargaining power, participation in decision-making, and access to inputs and services (Hossain and Jail, 2011). The majority of women are still denied their right to access and own essential resources, right to information, and freedom of action.

Further, there is evidence that the existing statutory and customary laws still restrict women's access to essential productive resources such as land and other types of property in most of the African and Asian countries (Nuria, 2016). Also, women continue to be in a disadvantaged position in many aspects though they contribute a great deal to the wellbeing of their societies as 79 percent of them engage in agricultural production (Adeyeye *et al.*, 2019). As such, women empowerment is a strategic target of many development agenda as it is believed to have the potential to improve gender equality and reduce or get rid of poverty (Adeyeye *et al.*, 2019; Galiè *et al.*, 2018). This explains why women empowerment has been taken further aboard in the 2030 agenda for sustainable development (UN, 2018; Lee and Finlay, 2017).

In developing countries, policy makers place women empowerment at the centre of development policy decisions. Likewise, donor agencies, local governments and NGOs are increasingly targeting women as a priority and are strengthening their investment in the efforts to empower women and reduce gender inequality (UN, 2015; Gates, 2014). It is to be borne in mind that empowerment is a multi-dimensional, dynamic and complex endeavour, involving changes in power relations so as expand individual choices and self-reliance (Kyamusugulwa, Hilhorst and Bergh, 2019). Women empowerment is defined as enhancing women's ability to make strategic life choices in a context where the freedom to do so is denied to them (Malhotra *et al.*, 2009);

Kabeer, 2005; Kabeer, 1999). It focuses on boosting the status of women at the individual, household level, community level and beyond (Bayeh, 2016; Malhotra *et al.*, 2009).

Various empowerment strategies have been implemented by development partners in different parts of the globe. Gender mainstreaming in development projects has become a common strategy for challenging women's subordinate position and oppression in their families and societies (Jeckoniah *et al.*, 2012). Other strategies focus on raising women's status through education, training, micro enterprise interventions, leadership training and quotas (Halim *et al.*, 2015; Goldman and Little, 2015) and improving their access to credit and microfinance programmes, health and family planning services as well as legal counselling and support (Jeckoniah *et al.*, 2012). There are also those which focus on empowering women by improving the agricultural-based sources of their livelihoods such as crop cultivation, forestry, fisheries, livestock keeping, and agro-industries (Galiè *et al.*, 2018).

Recently, interventions implemented under development projects have emerged as a more efficient and effective strategy for reducing women's poverty and, hence, empowering them (Tesfaye *et al.*, 2018; Johnson *et al.*, 2017; Chiarini, 2017). Most of these projects target poor rural communities where women constitute the majority of the population (Chiarini, 2017; Ntale and Litondo, 2013). Livestock-related interventions are also reported to play an important role in fostering gender equality (Galiè *et al.*, 2018). However, these interventions have received relatively little attention from researchers (FAO, 2014).

Notwithstanding the various efforts initiated in order to empower women, there is evidence indicating that quite a plenty of women still lack access to and control over resources such as land, capital, agricultural inputs and technology (Fletschner and Kennedy, 2014).As such, the thesis on which this book is based investigated the role of improved chicken intervention in empowering women socioeconomically in Bariadi and Muheza District in Simiyu and Tanga Regions respectively.

Bariadi and Muheza are among the districts where improved chicken interventions were introduced by the African Chicken Genetic Gain (ACGG) project. ACGG is a development partnership project implemented in Tanzania, Ethiopia and Nigeria in order to develop public-private partnerships so as to improve chicken productivity and promote women empowerment (ACGG, 2015). ACGG placed women at the centre of the project activities, consistent the argument by Galiè *et al.* (2018) that women should be the central focus if Agricultural Research for Development (AR4D) is to be successful. The ACGG project is unique since it engaged the private sector from the beginning and adopted innovation platforms through which stakeholders could identify challenges together and co-create solutions. The Innovation Platforms were meant to provide a basis for catalysing a private-sector-led transformation of the chicken value chain. In Tanzania, the project benefited 20 districts in five agro-ecological zones.

A number of improved chicken interventions have been initiated to improve the livelihoods of poor people and women in particular in rural Tanzania. This is associated with the recognition of development interventions as a potential strategy for empowering women in terms of income, food security and ownership of assets (Nordhagen and Klemm, 2018; Lyimo, 2013). For instance, a poultry-for-nutrition intervention was implemented by Helen Keller International (HKI) in four African countries, namely Burkina Faso, Tanzania, Senegal and Cote d'Ivoire from 2013 to 2016. The project was known as Creating Homestead Agriculture for Nutrition and Gender Equity Project (CHANGE). Among others, the project aimed to improve women's access to and control over productive resources and raise women's income through the sales of surplus production. In Tanzania, CHANGE was implemented in Sengerema and Ukerewe districts of Mwanza Region around Lake Victoria. HKI used a community-based approach to provide participants with extension services and start-up inputs in collaboration with local non-governmental organizations and government agencies (Nordhagen and Klemm, 2018).

In the recent years, rural poultry genetic improvement programmes have been geared towards adoption of improved chicken breeds that have high productivity, adaptability and resistance against disease (Wondmeneh *et al.*, 2014). In Tanzania, the African Chicken Genetic Gain (ACGG) project began in 2015 with the aim of empowering women by improving chicken productivity. However, evidence on whether specific agricultural development projects actually empower women is limited (Johnson *et al.*, 2017; Tesfaye *et al.*, 2018). Various scholars argue that projects that target women may increase their (women's) access to income but do not necessarily contribute to women empowerment (Johnson *et al.*, 2018; Santos *et al.*, 2014; Quisumbing *et al.*, 2013; FAO, 2011; USAID, 2009). The effects of specific interventions on women empowerment depend on the socio-economic and cultural set up, and thus it is difficult to make generalizations. In order to fill this knowledge gap, the current study assessed the contribution of improved chicken intervention projects to women's social-economic empowerment in selected districts. The study was based on the assumption that when women are economically and socially empowered, they become potential agents of change.

1.2   Theoretical Framework

The study was guided by Women Empowerment Theory and Innovation Systems Theory. Women Empowerment Theory articulates agency, structure and relations as important domains of empowerment in the context of improved chicken interventions. Innovation Systems Theory on the other hand stipulates that Community Innovation Platforms (CIPs) are essential in promoting women empowerment.

1.2.1 Women Empowerment Theory

Kabeer (2005) explains empowerment as a process that has three dimensions, namely agency, structures and relations. Agency refers to the processes by which choices are made and put into effect. A person or group's agency can largely be predicted by their possession of assets. Assets are the stocks of resources that enable actors to use economic, social and political opportunities to improve their lives (Alsop *et al.*, 2006). Structure comprises the institutions that govern people's behaviour and influence the success or failure of the choices that they make. Institutions can be formal or informal. Formal institutions include the rules, laws and regulatory frameworks that govern the political processes, public services, private organizations and markets. Informal institutions are "unofficial" rules that structure incentives and govern relationships within organizations, informal cultural practices, value systems and norms that operate in households or among social groups (Alsop *et al.*, 2006). Structure can have recognizable forms such as the form in which households are organised (monogamous, polygamous, etc.), producer groups, development agencies, government institutions and laws (Farnworth *et al.*, 2013).

Relations are closely linked to structure and agency. Relations are those connections that people have both within and outside their communities. Thus, relations are about women's freedom to participate in groups, to take part in coalitions to claim their rights to land and other resources, to be directly engaged in development and be acknowledged for their work, and to participatee actively in value chain partnerships such as in producer and marketing groups and in value chain platforms (Farnworth *et al.*, 2013). Relations make it possible for one to create, participate and benefit from networks or coalitions

As per the theoretical perspectives, it is assumed in this study that women empowerment can be achieved through an interaction of the three dimensions: women's agency, structure (cultural norms and beliefs) and relations (participation of women in groups, networks/innovation platforms). The study sought to understand whether improved chicken intervention improves agency and whether agency influences change in the structure and relations.

1.2.2 Innovation System Theory

Innovation System Theory emphasizes on the need for more communication and collaboration among heterogeneous groups and the harnessing of abilities (knowledge, skills, capabilities, interests, resources) to address the existing challenges to innovation and take advantage of potential opportunities (Sell *et al.*, 2018; Lundvall, 2010; Malhotra *et al.*, 2009). Innovation platforms (IP) are based on innovation systems thinking. They are considered as a holistic and comprehensive framework for understanding innovation. So, in agriculture, IPs can be useful in exploring the strategies that can boost productivity, sustainably manage natural resources and improve value chains or influence policies (Malhotra *et al.*, 2009).

In this study, Community Innovation Platforms (CIPs) provided a basis for understanding how diverse actors including women collaborate and co-create solutions to their common challenges along the chicken value chain and the community at large. Cultural norms and practices could be questioned through multi stakeholder collaborations. It is therefore assumed that effective engagement of diverse actors in Community Innovation Platforms is fundamental for realizing women empowerment.

1.3   Conceptual Framework

Women empowerment is an active, complex and multidimensional process which enables women to actualize their full potential and power in all spheres of life (Pan, 2017; Gupta, 2017; Mathialagan, 2015; Mathialagan, 2014). It is a complex process since it comprises economic, socio-cultural, familial/interpersonal, legal, political, psychological dimensions, etc. (Assaad *et al.*, 2014; Malhotra *et al.*, 2002). Being that complex, it is defined and interpreted differently from varied standpoints. There are several frameworks for evaluating and measuring empowerment, including those developed by Laven and Verhart (2011); Samman and Santos (2009); CARE (2009); Alsop *et al.* (2006), KIT *et al.* (2006); Kabeer (2005) and Alsop and Heinsohn (2005). The CARE Women Empowerment Framework focuses on amplifying women's voices and their participation from the household level to higher levels. Critically, the framework stipulates that women's participation will be effective only if efforts are made from the individual level to the structures and processes that affect women's access and control over assets of all kinds (structure) and the networks (or relations) that enable women to interact effectively with development actors. All these frameworks are referred to in the evaluation of agency, structure and relations.

Therefore, the study adopted an empowerment framework with three interrelated dimensions from Alsop *et al.* (2006) and CARE (2009) as illustrated in Figure 1.1.

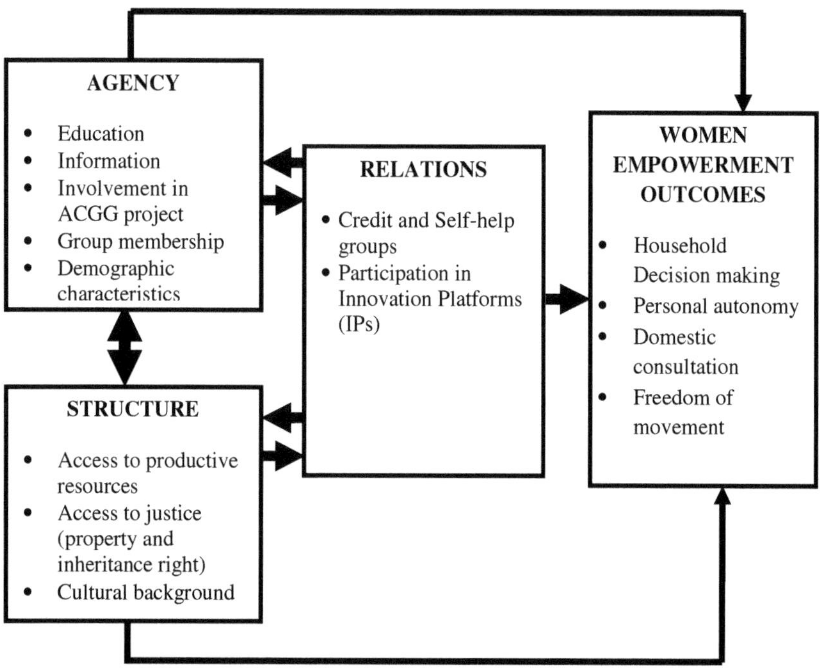

Figure 1.1: *Conceptual Framework adapted from Alsop et al. (2006) and CARE (2009)*

1.4 General Study Methodology

1.4.1 Study area

This study was conducted in Bariadi and Muheza districts in Simiyu and Tanga regions respectively. Bariadi and Muheza are among the districts that benefited from the ACGG project. Other districts that benefited from the ACGG project are: Masasi, Newala, Ruangwa, Lindi Rural, Mbeya Rural, Ileje, Njombe Rural, Wanging'ombe, Manyoni, Iramba, Bahi, Chamwino, Kilombero, Morogoro Rural, Korogwe, Maswa, Misungwi and Sengerema. Bariadi and Muheza districts were purposefully selected as the study area due to the socio-cultural and socio-economic differences between them. Bariadi District is predominantly inhabited by agro-pastoral Sukuma people, while Muheza District is inhabited by a large diversity of ethnic groups, most of which engage in agriculture. In each district, three villages were included. Specifically,

Mwamoto, Byuna and Ibulyu villages were selected in Bariadi District, while Kwaisaka, Kisiwani Nkumba and Mlingano villages were selected in Muheza District.

1.4.2 Research design

The study adopted a cross-sectional research design which enables the observation of two or more variables at a single point in time. The design is suitable for description purposes and for determining relationships among variables (Babbie, 1990). The design is commonly used in survey research to differentiate at least two categories of people (Malamsha, 2014). This study included two categories of women, namely beneficiaries of ACGG and non-beneficiaries.

1.4.3 Sampling procedure and sample size determination

The sampling unit of the study was a woman aged 18 years and above. The population of the study comprised both beneficiaries and non-beneficiaries of the African Chicken Genetic Gain (ACGG) intervention. The beneficiaries of ACGG intervention are individuals who were provided with improved chickens, whereby each participating household received 25 pre-brooded and pre-vaccinated chickens. A list of ACGG beneficiaries and a list of non-beneficiaries from village registers were used as sampling frames. Simple random sampling using the lottery method was employed in the selection of individuals from the list of names from each village. The sample size was 240 women (120 beneficiaries and 120 non-beneficiaries from Bariadi and Muheza Districts respectively). This samples size was determined by using Cochran's formula (1977). $n = {^N}/{1 + N\,(e)^2}$ Where: n = Sample size; N = the population size; e = the level of precision or sampling error, estimated in percentages (0.05).

1.4.4 Data collection methods

Mixed methods of data collection were used to gather quantitative and qualitative data we. The use of mixed methods was preferred in order to ensure validity and reliability the data through triangulation (Creswell, 2013). A structured questionnaire was used to gather quantitative data on household socio-economic characteristics, ownership of assets, access to the means of production, and the socio-cultural settings as related to keeping of improved chicken breeds. Key informant interviews and Focus Group Discussions (FGDs) were used to collect qualitative data. The qualitative data on women empowerment were collected from May to June, 2018 using FGDs and key informant interviews. Focus Group Discussions involved women and men who were engaging in keeping improved chickens and those who had received training on management of improved chickens. The issues discussed in the FGDs included the role of stakeholders in the chicken value chain, that is, from access to the productive resources all the way through production and marketing of the chicken products up to the consumption of those products. A total of twelve FGDs were conducted in six villages. i.e. two in each village. The number of participants per FGD ranged from 8 to 10, which is in line with Barbour (2011) and Bryman (2004) who argue that, when the FGD participants are too many, some of them just sit idle without sharing any opinion but and when they are too few, they may not be able to discuss difficult topics effectively.

Key informant interviews were held with people who were considered to have in-depth knowledge and understanding on chicken production and the role of improved chicken keeping in promoting women empowerment. The key informants included 2 District Livestock and Fisheries Officers (DLFOs), 2 District Community Development Officer (DCO), 6 Village Extension Officers (VEO) and 2 ACGG zone coordinators. The key informant interviews were carried out to collect views from leaders and individuals on women's issues in the context of improved chicken keeping.

1.4.5 Data analysis

In respect to the first objective of the study, qualitative data were analysed using content analysis procedures after organising them into different themes that addressed the objectives. Quantitative data were analysed by using the Ordinal Regression Model and descriptive statistics. Women empowerment was measured by using a Women Empowerment Index (WEI). Four women empowerment indices were developed and used to derive a composite empowerment index (CEI). The composite empowerment index was derived by averaging four women empowerment indices, which included Household Decision-Making Index (HDMI), Domestic Consultation Index (DCI), Personal Autonomy Index and Freedom of Movement Index (FMI). HDMI was used to know who makes decisions on day-to-day expenditures, purchases of assets, use of family income, sales of chicken and eggs and on construction chicken house. A score of 1 was given if a woman alone made decisions, a score of 0.5 was given if both husband and wife made decisions and a score of 0 was given if only the husband made decisions. The Personal Autonomy Index was used to determine whether a woman can attend training, visit the market or plan to use money without asking for permission from her husband. A weight of 1 was given if the woman asked for permission frequently, 0.5 if the woman occasionally asked for permission and 0 if the woman never asked for permission. The Domestic Consultation Index was used to find out whether a woman was consulted by her husband when he wanted to expend the income obtained from chicken sales or to do any other development activity. A score of 1 was given for frequent consultation, 0.5 for occasional consultation and 0 for no consultation. The Freedom of Movement Index was determined to know whether a woman had freedom to visit the market, attend meetings and engage in social activities. A score of 1 was used for frequent freedom, 0.5 for occasional freedom and 0 for no response. The average of these four women empowerment indices resulted to Composite empowerment index (CEI), which was broken into four levels of empowerment, "No empowerment" for the average of 0, "low empowerment" for the average range from 0.1 to 0.5, "medium empowerment" for the average range from 0.6 to 0.7 and "high empowerment" for the average score from 0.8 and above. The Human Development Index developed by UNDP (2005) was adopted in categorizing the empowerment levels. CEI was computed by averaging the four indices.

$$Y = 1/4\ (HDM+DCI+FMI+PAI) \dots\dots\dots\dots\dots\dots\dots\dots\dots\dots\dots\dots\dots\dots\dots\ (1)$$

UNDP's classification of the human development index was adopted, where empowerment was chunked into four levels. A score of 0 on the composite empowerment index was categorized as "No empowerment", scores of 0.1 – 0.5 were categorized as "low empowerment", scores of 0.6 – 0.7 were categorized as "medium empowerment" and a score higher than (0.8) was classified as "high empowerment". Various scholars such as Jeckoniah *et al.* (2012); Alam *et al.* (2015) and Sheikh *et al.* (2016) have also ever used CEI as a measurement of women empowerment.

The socio-economic factors that influence women empowerment were determined by using the Ordinal Logistic Regression Model. The model was relevant because the dependent variable (Y) was classified in terms of ordered empowerment levels (low, medium and high). As per Agrest and Finlay (2009), ordinal logistic regression is appropriate when the outcome is at an ordinal level with more than two categories.

The model is defined as follows:

$$P\ (Y) = \frac{e^{\alpha + \beta1x1 + \dots \beta kxkk}}{1 + e^{\alpha + \beta1x1 + \dots\ \beta kxk}} \dots\dots\dots\dots\dots\dots\dots\dots\dots\dots\dots\dots\dots\dots\dots\ (2)$$

Where:

$P(Y)$ = the probability of the success alternative occurring, Y = dependent variable, e = the natural log, α = the intercept of the equation, β_1 to β_k = coefficient of the predictor variables, X1 to X_k = independent variables entered in the regression model.

The dependent variable Y represented three levels of empowerment, namely: low empowerment, medium empowerment, and high empowerment. The independent variables Xs are socio-economic characteristics which, according to various theories and empirical evidence, are the factors that influence women empowerment. These included age, family size, education, religion, marriage type, engagement in ACGG interventions, membership to an organization, and membership to the Community Health Insurance Funds (CHIF).

The model is further expressed as:

$$Y = \alpha + \beta_1 X_1 + \beta_2 X_2 + \beta_3 X_3 + \beta_4 X_4 + \beta_5 X_5 + \beta_6 X_6 + \beta_7 X_7 + u \dots\dots\dots\dots\dots\dots\dots\dots\dots\dots(3)$$

Where by

Y = Levels of empowerment (CEI)

α = constant of the equation

β_1 - β_5 = the coefficient of the independent variables that represent a unit change in dependent variable due to a change in independent variable.

X_1 = Age of respondents (expressed in years)

X_2 = Family size (expressed in number of people)

X_3 = Education level of the respondent (expressed in years of schooling)

X_4 = Type of marriage (Monogamy = 0, Polygamy =1)

X_5 = Involvement in ACGG project (Non ACGG=0, ACGG=1)

X_6 = CHIF member (Non-member =0, Member =1)

X_7 = Group membership (Non-member = 0, Member =1)

X_8 = Religion (Muslims=0, Christians = 1)

U = Error term

The data for the second and third objectives were analysed by using content analysis procedures and descriptive statistics.

2.0 SOCIO-ECONOMIC DETERMINANTS OF EMPOWERMENT AMONG WOMEN PRODUCERS OF IMPROVED CHICKENS IN BARIADI AND MUHEZA DISTRICTS, TANZANIA

2.1 Background information

There has been an emerging consensus within the international development community that gender equality and women empowerment are important goals from the human rights perspective and in achieving a range of economic and social development objectives (Assaad *et al.*, 2014; Johnson *et al.*, 2017). Women empowerment occupies a vital position among the 17 global sustainable development goals and the 169 indicators adopted by the UN General Assembly in 2015 as of the 2030 agenda for sustainable development (Lee and Finlay, 2017). The issue of women empowerment gained prominence during the Beijing Conference of 1995 (UN, 2015). Yet, twenty years after the Fourth World Conference on Women, no country has fully achieved gender equality, and empowerment of women and girls (van Eerdewijk *et al.*, 2017; UN, 2015; UN, 2014).

Women empowerment is an active, complex and multidimensional process intended to enable women to realize their full potentials and power in all spheres of life (Pan, 2017; Gupta, 2017; Mathialagan, 2015; Mathialagan, 2014). It has numerous dimensions, including economic, socio-cultural, familial/interpersonal, legal, political and psychological dimensions (Assaad *et al.*, 2014; Malhotra *et al.*, 2002). It is a complex concept that is usually defined and interpreted differently from diverse standpoints. In this paper, it is considered as a process geared towards making women have access to and control over economic resources, and ensuring they can use them to exert increased control over other areas of their lives (Hunt and Samman, 2016; Taylor and Pereznieto, 2014). So, the focus of the paper is its socio-economic dimensions.

The conceptualization of women empowerment is mainly based on three defining elements namely agency, resources and achievement, which are common in the existing empowerment frameworks (Kabeer, 2005). The first defining feature is agency, which is the "ability to define one's goals and act upon them" (Kabeer, 1999) or the ability to control various aspects of one's life (Kishor and Gupta, 2004). The second element is access to and control over resources (materials, human and social), which a woman acquires from her relationships in the various domains such as family, market and community. The third element is the broader setting that characterizes the circumstances of a woman's life, such as marriage, living arrangements, household wealth and characteristics of the influential family members, opportunities and choices available to her (Kabeer, 1999). Throughout the world, women empowerment has increasingly become a concern among development partners such as governments, NGOs and development organizations. Various efforts to empower women at the household level and beyond have been initiated. These efforts have mostly been focusing on raising women's status through education, and training (Hunt and Samman, 2016; Aslam, 2013; Jeckoniah *et al.*, 2012). Other efforts have been focusing on increasing women's access to health and family planning services, representation in decision-making organs, involvement in credit and microfinance programmes, and access to legal counselling and support services (Kumar, 2015; Jeckoniah *et al.*, 2012).

Given the importance of empowering women, development practitioners now incorporate the women empowerment objectives into the design and implementation of their agricultural projects and programmes (Johnson *et al.*, 2017; Jeckoniah *et al.*, 2012). For instance, in the livestock sector, the African Chicken Genetic Gain (ACGG) project in Tanzania has introduced genetically improved chicken breeds in rural areas with the aim of increasing smallholder chicken productivity and promoting women empowerment (ACGG, 2015). Due to the complexity and multidimensional nature of women empowerment, various factors could differently influence the extent to which women are empowered in different contexts (Akter *et al.*, 2017; Assaad *et al.*, 2014). The relative importance of each factor differs from dimension to dimension and from context to context (Samman and Santos, 2009).

Therefore, this paper examines the determinants of women empowerment in the context of improved chicken production in selected districts. The African Chicken Genetic Gain (ACGG) project was implemented in five agro-ecological zones of Tanzania (ACGG, 2015). The project put women at its centre and distributed improved chicken breeds to them to test their acceptability and ultimately establish the extent to which the adoption of these breeds would impact women's life.

2.2   Socio-economic profile of the respondents

The respondents (both ACGG beneficiaries and non-beneficiaries) had a mean age of 42 years, which is a productive and active working age. With this mean age, the selected women were the best representative sample since most of them were experienced with issues related to decision-making in the household and were engaging in various economic activities. Details about the age of the respondents are shown in

Table 2.1. In terms of education, the majority of the respondents, (69.2% and 72.5% of the ACGG beneficiaries and non-beneficiaries respectively) had attained primary education (see Table 2.1). With such level of literacy, they were considered to have enough access to information, which they could be integrating into various economic activities.

Table 2.1: Distribution of respondents by socio-economic characteristics (n=240)

Variable	ACGG Beneficiaries (n=120)		Non-Beneficiaries (n=120)		All	
	Frequency	%	Frequency	%	Frequency	%
Age						
15-24	2	1.7	6	5	8	3.3
25-54	92	76.7	91	75.8	183	76.3
55-64	16	13.3	20	16.7	36	15
65 and Above	10	8.3	3	2.5	13	5.4
Education						
No formal education	28	23.3	27	22.5	55	22.9
Primary	83	69.2	87	72.5	170	70.8
O-level	9	7.5	5	4.2	14	5.8
A-level	0	0	1	0.8	1	0.4
Marital status						
Single	12	10.0	9	7.5	21	8.8
Married	98	81.7	97	80.8	195	81.3
Separated	3	2.5	4	3.4	7	2.9
Widow	7	5.8	10	8.3	17	7.0
Forms of Marriage						
Monogamy	68	69.4	73	75.3	141	72.3
Polygamy	30	30.6	24	24.7	54	27.7
Religion						
Christian	56	46.7	55	45.8	111	46.3
Muslim	36	30.0	39	32.5	75	31.3
None	28	23.3	26	21.7	54	22.5
Economic activity						
Crops and livestock	100	83.3	102	85.0	202	84.2
Crops, livestock and petty trade	20	16.7	18	15.0	38	15.8

The results further showed that over 80% of the respondents were married and monogamy was the commonest form of their marriages, with 72% of them falling into this category. All respondents depended on agriculture (crop production and livestock keeping) as their main source of livelihood.

2.3   Women empowerment levels and socio-economic factors

The statuses of women empowerment based on socio-economic variables are shown in Table 2.2. The findings revealed that, overall, women in the study area had a medium level of empowerment (CEI = 0.622). Women who engaged in keeping improved chicken breeds had a higher empowerment level (CEI = 0.714) compared to those who did not (CEI = 0.529). The CEI of the ACGG beneficiaries was statistically higher than that of non-beneficiaries (F = 40.20, p < 0.001). In Bariadi, the level of women empowerment was a little higher (CEI = 0.648) than that of women in Muheza District (CEI = 0.595). This is because Bariadi is dominantly inhabited agro-pastoral Sukuma people whereby women mostly engage in poultry while men dominantly engage in keeping large animals like cattle. Therefore, decisions regarding chicken keeping were mainly made by women. It was also found that the empowerment level of widowed, separated and single women was higher (CEI = 1.000) than that of married women (CEI = 0.534). In terms of the type of marriage, women who were in polygamous marriages had a higher level of empowerment (CEI = 0.617) compared to women in monogamous marriages (CEI = 0.503). The findings indicated that women in polygamous marriages spent most of their time alone and hence they were not directly affected by the patriarchal system at the household level. Most of the decisions about their lives were made by them, something that raised their empowerment level.

In Muheza district, the empowerment level was found to increase with an increase in the education level of the respondents. This was not the case in Bariadi District, perhaps because a big number of women did not have formal education. According to the findings, 38.3% of women in Bariadi had no formal education while only 7.5% of the respondents had no formal education in Muheza. In this case, the tacit knowledge accumulated through life experiences might have played a significant role in empowering the Bariadi women. Furthermore, the level of women empowerment was found to be increasing with age for all groups of respondents. Meanwhile, Christian women (CEI = 0.6170) were found to be more empowered than Muslim women (CEI = 0.5787), something that is attributable to the fact that patriarchy is stronger among Muslims than among Christians.

Table 2.2: Status of Women Empowerment by Selected Socio-Economic Variables

Socio-demographic variable	Mean index		
	Muheza	Bariadi	Muheza & Bariadi
1. District	0.595	0.648	0.622
2. Participation			
-Women with ACGG chicken	0.685	0.743	0.714
-Women without ACGG chicken	0.505	0.553	0.529
3. Marital status			
-Single	1.000	1.000	1.000
-Married	0.466	0.594	0.534
-Separated	1.000	1.000	1.000
-Widow	1.000	1.000	1.000
4. Marriage(Form)			
-Polygamy	0.586	0.621	0.617
-Monogamy	0.456	0.572	0.503
5. Wives reside the same compound with the husband			
- No	0.600	0.657	0.648
- Yes	0.458	0.571	0.511
6. Education status			
- No formal education	0.522	0.728	0.695
- Primary education	0.604	0.589	0.598
- Secondary education	0.575	0.833	0.627
7. Age group			
15-24	0.625	0.350	0.488
25-54	0.557	0.629	0.594
55-64	0.650	0.738	0.689
65 and Above	0.900	0.917	0.908
8. Economic activities			
-Farming and livestock keeper	0.623	0.647	0.635
-Farming, livestock and petty trade	0.447	0.658	0.553
9. Religious			
-Christian	0.615	0.608	0.611
-Muslim	0.582	-	0.582
-Paganism	-	0.698	0.698

2.4   Determinants of women empowerment

The results of ordinal logistic regression model (Table 2.3) revealed that age, religion, type of marriage, and involvement in the ACGG project significantly influenced women empowerment in Bariadi and Muheza District (p < 0.05). The age of the respondents had a positive coefficient, which means that it significantly influenced women empowerment. The positive coefficient of age implies that women empowerment increases with increase in age. As women grow older, the level of respect accorded to them increases following the recognition of their contribution in the household. Therefore, older women gain confidence in decision-making at the household level, high freedom of movement and high personal autonomy. This is consistent with reports by Heshmati (2017); Nyange *et al.* (2017) and Peterman *et al.* (2015). However, it disagrees with Jeckoniah *et al.* (2012) who found that younger women and those aged above 50 years had lower levels of empowerment. In the current study, women beyond 50 years of age were found having a high empowerment level. This implies that the influence of age on women empowerment varies depending on the context and nature of intervention.

Involvement of women in ACGG had a negative coefficient and strong influence (Wald = 76.88) on women empowerment (Table 2.3). This implies that women who were not involved in the ACGG intervention were relatively less empowered than their counterparts who were involved in that project. Involvement in the ACGG project enabled women to access resources, training and advice related to the production of improved chickens and gender relations. This finding was affirmed by the following information from a key informant:

> Women who are engaged in the ACGG project are more empowered than those who are not engaged in it. The access to productive resources of the improved chickens enables them to benefit from sales of eggs and cocks. Therefore, women's earnings from improved chickens provide opportunities for women to contribute to the households as well as women's involvement in decision making (Extension officer at Kwaisaka village, 10.05.2018).

The findings are similar to those Haghighat (2014) which showed that access to resources puts women in a position to advance their social status and power. Accordingly, Highighat study showed that, apart from giving women access to productive resources, the ACGG project set conducive environment for stakeholder engagement, particularly in challenging the existing social structures that hinder women empowerment.

Table 2.3: Factors influencing women empowerment (n=240)

Variables	Coefficient	S.E	Wald	Sig.	95% C.I	
					L.B	U.B
Age	.048	.019	6.373	.012	.011	.086
Family size	-.042	.071	.343	.558	-.182	.098
Education	.067	.065	1.051	.305	-.061	.194
Type of marriage	-1.481	.462	10.264	.001*	-2.387	-.575
ACGG beneficiary	-4.598	.524	76.883	.000*	-5.625	-3.570
CHIF member	.051	.399	.016	.899	-.731	.832
Organizational member	-.297	.374	.630	.427	-1.030	.436
Religion	-1.956	.498	15.451	.000*	-2.931	-.981

P = 0.000; Goodness of Fit=1; Cox and Snell R^2= 0.590; Nagelkerke R^2= 0.677; Test of Parallel line = 0.849

The findings of the study further revealed a significant variation in women empowerment based on religions (p < 0.01). Christians were more empowered than Muslims. Hence, since Bariadi District had more Christians (66%), women in that district were more empowered than women in Muheza where Muslims comprised 62% (Table 4). Christianity provides opportunities for women to have a certain power of control over household assets but Islam ascribes man as the overall controller of household resources and benefits. The results are in line with Njoh and Akiwumi (2012) who report that Christianity has a more positive influence on women empowerment than Islam. Christianity emphasizes collaboration between women and men on the issues related with access to and control over household resources. However, this depends on the extent to which one is committed to the Christian teachings.

Moreover, the results indicated that a considerable number of Sukuma people in Bariadi are basically Christians strongly holding some cultural norms and beliefs which place women and girls in a subservient position. Regarding this, one of the key informants said:

> ...the traditional dance known as mbina promotes early marriage whereby girls are denied the right to education (Extension Officer, Bariadi District, 05.06.2018).

Also another key informant reported that:

> ...girls are advised not to study hard in school by their mothers. When it happens that a girl has passed primary school final examination, her mother is likely to

be divorced. Girls are expected to get married soon after primary school so as to bring wealth in terms of dowry price to their fathers (Village Extension Officer, Mwamoto Village, 01.06.2018).

The results are in line with James (2018) who found that religious beliefs are among the causes of discriminatory practices against women. The results imply that social institutions, including most of religious beliefs and practices, continue to restrict women from accessing productive resources, income and education, something which constrains the efforts to empower women. Efforts towards changing the unfavourable institutional settings (structure) and religious beliefs in particular could enhance women empowerment.

Table 2.4:   ***Distribution of respondents by religious affiliations (n=240)***

	Muheza			Bariadi		
Religion	**Frequency**	**%**		**Religion**	**Frequency**	**%**
Christian	46	38.3		Christian	79	65.8
Muslim	74	61.7		Muslim	1	0.8
Traditional beliefs	0	0.0		Traditional beliefs	40	33.3
Total	**120**	**100.0**		**Total**	**120**	**100.0**

Moreover, there was significant variation in women empowerment based on the types of marriage ($p < 0.01$). In a monogamous system, men's over control of resources is stronger than in a polygamous system whereby one woman may be left alone to make her own decisions. The findings indicated that Muheza District, which has a high proportion of Muslims, is dominated by a monogamous marriage system. This is inconsistent with Newbury (2017) who shows that polygamy is more common in Islamic societies. The difference could be due to location reasons and socio-economic forces that have changed people's attitude towards small manageable families.

Women who were in polygamous marriage were more empowered than those in monogamous marriages. In polygamous societies, women who do not reside in the same compounds with their husbands have greater chances of being more empowered. They make their own decisions, especially when the husband is absent or is elsewhere with another wife. This was affirmed by one of the women respondents in the FGDs at Ibulyu village in Bariadi who said:

> My husband has a second wife. Usually, he spends most of the time with the other wife. Therefore, I am responsible for making most of the decisions for the

household including production of improved chickens (ACGG woman beneficiary, 26.05.2018).

Similar results were obtained by Newbury (2017) who noted that most of the women in polygamous marriages exercise have more control over the use of resources generated from small income generating activities. However, due to patriarchal dominance in both Bariadi and Muheza, the level of autonomy that women have is limited. Overall, productive resources and the benefits accrued from income-generating activities continue to be under men's control.

2.5   Conclusions and Recommendations

Since women empowerment is multidimensional, the factors influencing it are context-specific. The findings of this study revealed that economic and socio-cultural variables, particularly age, religion, marriage type, and involvement in ACGG project have a significant influence on women empowerment.

The findings have indicated that Bariadi district had a higher illiteracy level than Muheza District. Surprisingly, while it is expected that education would result in higher levels of empowerment, Bariadi women's empowerment level was a little higher than that of Muheza women although the latter were more literate than the former. This calls for another variable to explain the phenomenon. Further, polygamy was found to be more common in Bariadi District than in Muheza District where Islamic religion, which allows polygamy, is dominant. Monogamy was found to be negatively associated with women empowerment as compared to polygamy in this context. This is because men have more control of household resources in the monogamous marriages than in polygamous marriages.

Involvement of women in the ACGG project sets conducive environment for women to have access to productive resources and for the stakeholders to challenge the cultural values that conflict with women empowerment. In this study, the findings indicated that women who participated in the ACGG project had a higher empowerment level than those who did not.

Based on the results, it is recommended that development stakeholders, including Local Government Authorities, Non-Governmental Organizations, Faith-Based Organizations, private sectors, donors and individuals, should implement similar interventions in other areas. Also, they should initiate educational programmes intended to raise awareness and get rid of harmful cultural norms and beliefs which hinder women empowerment.

The level of women empowerment in Bariadi District could be further improved if women have more access to continuing education. Education will increase their access to information on how to address traditional practices which impede women empowerment. In Muheza District, improved access to education could make them have access to information on the negative religious practices that contribute to the subordination of women and how to address them so as to enhance women empowerment.

3.0 THE CONTRIBUTION OF AFRICAN CHICKEN GENETIC GAIN (ACGG) PROJECT IN PROMOTING WOMEN EMPOWERMENT IN BARIADI AND MUHEZA DISTRICTS, TANZANIA

3.1 The role of the ACGG project

The ACGG project introduced genetically improved chicken breeds in rural areas with the aim of increasing smallholder chicken productivity and promoting women empowerment. The ACGG project provided women with inputs (initial stock of chicks and vaccines), training and extension services. It also organized local markets for them. The details on how ACGG intervention provided opportunities for chicken productivity and women empowerment are discussed below.

3.1.1 Provision of initial stock of chicks

During the in-depth discussion with the Eastern Zonal coordinator of the ACGG project, it was revealed that each woman beneficiary was given 25 chicks that were 6 weeks old. This was done to enable them to have access to productive resources, which is one of the factors that contribute to women empowerment. Similar results have been reported by Meinzen-Dick *et al.* (2019) and Haghighat (2014) who showed that access to productive resources is associated with women empowerment. Yet, even if women have access to productive resources, realization of empowerment objectives would depend on their commitment in utilizing the resources.

Moreover, the influence of resources on women empowerment depends on how such resources are organised, mobilised and managed. The way resources were organized, mobilized and managed in the two districts had faults in some sense. For example, the ACGG zone coordinators reported that caring of one day old chicks for six weeks was done by two brooders in both Bariadi and Muheza districts. One of the brooders from Muheza District had this to say:

> Brooding is a very sensitive activity which requires maximum care for the chicks particularly in feeding and vaccination. Once farmers received chicks from the brooders, they were required to vaccinate them after every three months starting from the day they received them from the brooder (Key informant, Mlingano village, 15.05.2018).

This excerpt shows that taking care of one day old chicks for six weeks was difficult for inexperienced persons due to a lack of facilities and inadequate skills on how to take care of the chicks. The results support Leroy *et al.* (2015) who assert that chicks should be raised for about 2 to 4 weeks and get vaccinated before being sold to the farmers to avoid juvenile mortality. Despite being taken care of by the brooders for six weeks, some of the chicks died soon after being handed to the farmers mainly due to stress caused by long distance transportation. In some cases, women's benefits from the project intervention were somehow affected by poor management practices as the women were not used to the appropriate management of the improved chickens.

3.1.2 Improved access to extension services and training

The ACGG project played a significant role in offering extension and training services to the beneficiaries. Access to extension services and training became more regular after the introduction of the project. However, the training package included information related to management of improved chickens rather than women empowerment issues. Similar findings have been reported by Galiè and Kantor (2016) who found that projects aimed at empowering women pay more attention to technical issues than to the socio-cultural settings that influence women's status. Equally, Cornwall (2016) revealed that providing women with the means to generate income alone cannot eliminate the deep-rooted structures that underlie gender inequality. Therefore, development projects aimed at empowering women should go beyond setting an enabling environment for women to access resources and opportunities for generating income. They should also focus on challenging the cultural norms and women's bad self-image that perpetuate gender inequalities.

3.1.3 Enhanced marketing of chicken and eggs

The findings revealed that selling of chickens and eggs was mainly organised at local markets which were not always reliable. Similar results have been reported by Thieme *at al.* (2014) that access to market is limited in rural areas. However, participants in FGDs at Ibulyu village in Bariadi District reported that, despite the market constraints, the ACGG project enabled women to organize themselves into marketing groups. They established a meeting centre for collecting and selling eggs. One of the participants in the FGDs said:

> Rearing of improved chicken breeds is among the potential sources of income for women. Every week I am sure to get at least Tsh. 6,000 from selling eggs (Woman ACGG beneficiary, 25.05.2018)

This excerpt implies that the organization of women into marketing groups made them have access to income as it assured them of a readily available market for their chicken products. However, the price of eggs at the farm gate was very low (Tsh. 300 per egg) as compared to the price of Tsh. 500 per egg in town centres. Therefore, strengthening the marketing groups could play a significant role in improving access to the market for chicken products and the prices of those products in particular. Also, improved access to market could increase the sustainability of improved chickens keeping as it is associated with costs of feeds, vaccines and shelter.

3.2   Respondent's opinions on the benefits of improved chicken

3.2.1 Income

The results show that a mean income of Tsh. 478 651 per household per year was obtained from selling eggs and chickens (Figure 3.1). One of the key informants reported that:

> The mean income was less than the expected income (Tsh. 800 000 to 1 000 000) due to limited markets, improper feeding and inappropriate vaccination. Some of the ACGG beneficiaries did not benefit from the improved chickens at all due to high chicken mortality (Extension Officer Muheza, 05.06.2018).

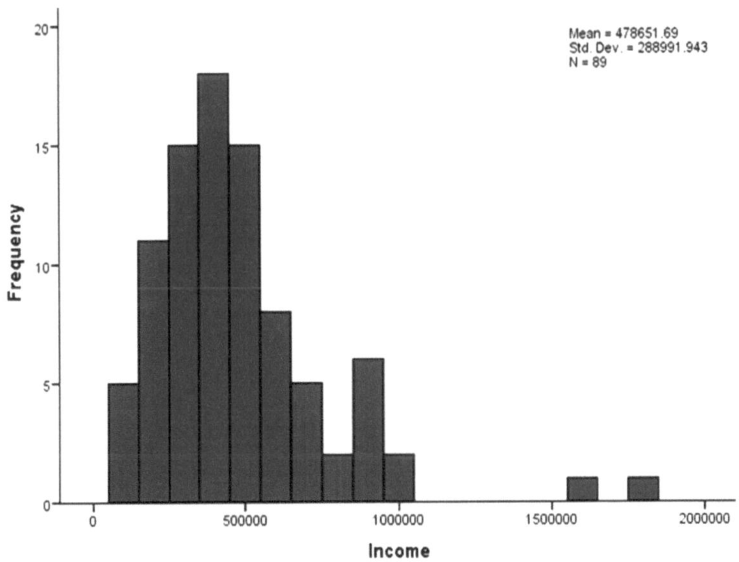

Figure 3.1:   *Distribution of respondents by income accrued from selling eggs and chickens*

The results imply that women's access to income improved as they reached at least half of the expected income. The income obtained from selling of cocks and eggs was spent on household needs. The majority of women who participated in FGDs reported that the income they earned from their improved chicken products was spent mainly on household necessities.

Some women used the benefits accrued from the sales of cocks and eggs to renovate or build huts for the chickens. This is unlike what studies such as Okitoi *et al.* (2007); FAO (2014) and Akite *et al.* (2018) revealed. These studies indicated that construction of the chicken huts is solely men's role. The findings of the current study thus imply that when women have access to income, they can effectively engage in activities that are traditionally done by men.

Following the increase in women's access to income, some men were worried that women behaviour would change. The majority of male FGD participants at Mwamoto village argued that women's access to more income might make them disrespectful to their husbands. This implies that, on seeing their wives having more access to income, men might develop mechanisms to limit that access to income. Regarding this, one of the women participants said in an FGD that, *men reduced their financial support when they realised increased women's access to income.* The results are in line with Santos *et al.* (2014) and Quisumbing *et al.* (2013) who revealed that women's projects may increase women's access to income, but may not necessarily promote women empowerment. Increased women's access to income may provide an opportunity for them to contribute to the household income that could be expended on necessities. However, the overall decision making power at the household level remains to be male-dominated. As a result, women's agency is constrained by the social-cultural settings which limit women's motivation to actively engage in keeping improved chickens.

3.2.2 Food availability

Results showed that improved chickens contributed to the food security at the household level (Table 3.2). On one hand, the income earned from sales of chicken products was used to buy food items. On the other hand, the chicken products were consumed at the household level. For instance, 35.0 % of the ACGG beneficiaries had schedules for eating eggs as compared to 13.3% of the non-beneficiaries who ate eggs (Table 3.2). The majority of the ACGG beneficiaries at Kwaisaka reported that they were eating eggs as part of their breakfast and for making lunch and dinner stew. Consumption of chicken meat was less common as the majority preferred to sell the chickens and the money from the sales to buy beef and other household necessities.

Table 3.1: *Distribution of respondents by egg consumption among ACGG and non ACGG respondents*

ACGG beneficiaries			Non- Beneficiaries		
Schedule	**Frequency**	**%**	**Schedule**	**Frequency**	**%**
Once a week	31	25.8	Once a week	8	6.7
Twice a week	7	5.8	Twice a week	4	3.3
Once after two weeks	4	3.3	Once after two weeks	4	3.3
Unplanned	78	65.0	Unplanned	104	86.7
Total	**120**	**100.0**	**Total**	**120**	**100.0**

According to the extension officer from Kwaisaka village, ACGG gave women access to eggs which in turn improved egg consumption in their households. Similar results have been reported by various studies such as Bashir and Schilizzi (2013) and de Bruyn (2018) which established that input availability is one of the most important determinants of food availability at the household level in the African context. Therefore, the ACGG project, through provision of initial chicks, extension and veterinary services, has played a significant role in increasing the availability of food in households.

3.2.3 Levels of women empowerment among ACGG beneficiaries and non-beneficiaries

The findings showed that 46.7% of the ACGG beneficiaries and 5.0% of non-beneficiaries had a medium level of empowerment while 21.7% and 72.5% of the beneficiaries and non-beneficiaries respectively had low levels of empowerment (see Figure 4.1). Higher level of empowerment for the beneficiaries was expected because ACGG provided women with access to productive resources through multi stakeholder collaboration. The results are in line with Meinzen-Dick *et al.* (2019) and Cornwall (2016) who report that women's access to economic resources enables them to make changes towards empowerment.

Apart from the support of productive resources, the ACGG project provided women with an enabling environment for stakeholder networking and raising of awareness on the social-cultural norms and practices that undermine women's status through trainings and Community Innovation Platforms (CIPs). Therefore, as reflected from the Women Empowerment Theory, the ACGG beneficiaries were empowered through interrelations of the three empowerment dimensions namely agency, structure and relations. However, the ACGG project focused more on provision of technical support related with keeping of improved chickens as it is a new innovation in the study area.

Awareness creation on the impact of cultural norms towards women empowerment was constrained with limited gender expertise, particularly at the local level.

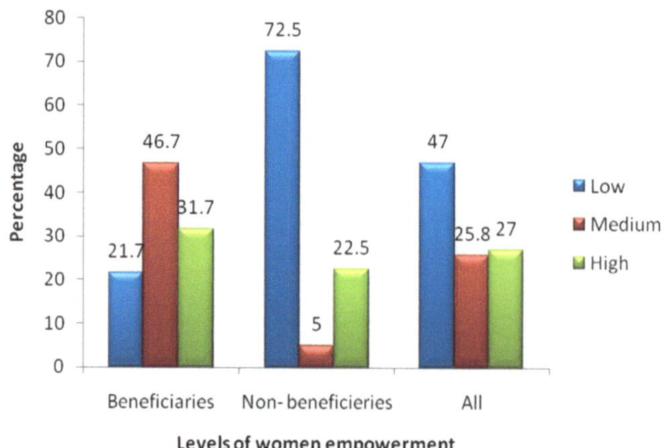

Figure 3.2: ***Distribution of women empowerment levels among ACGG beneficiaries and non-beneficiaries***

3.3 Conclusions and Recommendations

3.4.1 Conclusions

Improved chicken interventions for women provide potential opportunities for improving their livelihoods, access to resources and control over benefits. The African Chicken Genetic Gain (ACGG) project provided empirical evidence that improved chicken interventions can empower women to a great extent.

Social-economic empowerment entails both increasing women's access to income and boosting their ability to challenge cultural norms and practices that influence power relations. The ACGG project enabled women' to have access to productive resources of improved chicken through multi-stakeholder collaboration. Community innovation platforms were established to strengthen multi-stakeholder collaboration towards addressing innovation challenges in the chicken value chain. Among others, women were encouraged to effectively participate in innovation platforms.

Moreover, keeping of improved chicken is a new innovation in the study area. Therefore, changing women's mind-sets from keeping indigenous breeds to keeping improved breeds is a slow process that requires women to have entrepreneurship skills and understand what high inputs/high outputs entail in relation to the activity of keeping improved chicken breeds.

3.4.2 Recommendations

It is recommended that the ACGG project, Local Government Authorities and development partners have to extend the improved chicken interventions to other areas. Moreover, the LGA through the departments of extension, livestock and Community Development in collaboration with other development partners should enhance their services and integrate women empowerment objectives in improving chicken productivity and development projects at large.

Training institutions, the Local Government Authority (LGA), Community-Based Organizations (CBOs), Faith Based Organizations (FBOs), Non-Governmental Organizations (NGOs) and the private sector should provide training   on the influence of cultural beliefs and norms that impede the efforts of empowering women. This can be done through training, awareness raising, innovation platforms and lobbing and advocacy.

4.0 THE ROLE OF COMMUNITY INNOVATION PLATFORMS IN PROMOTING WOMEN EMPOWERMENT IN BARIADI AND MUHEZA DISTRICTS, TANZANIA

4.1 Background information

Being the main source of livelihoods for the rural poor, agriculture is in the target of the efforts for attainment of sustainable development goals. In sub-Saharan Africa, agriculture is dominated by small scale farmers and women in particular (Sell *et al.*, 2018). While women are the main producers in agriculture, their contribution remains largely unrecognised (Diiro *et al.*, 2018; World Bank, 2009). At the same time, gender inequalities that put women in a disadvantaged position limit agricultural productivity and efficiency, hence retarding development. Recently, agricultural development has undergone transformation as a result of multiple approaches, beginning with the linear technology delivery model to a less linear knowledge exchange model characterised by a feedback system, and eventually to the Agricultural Innovation Systems (AIS), which integrates all actors equally in the innovation platforms (Mbo'o-Tchouawou *et al.*, 2016). Agricultural Innovation Systems (AIS) stresses the need for closer linkage between the initiators of innovation (technological actors) and the promoters/users (Socio-economic actors) either in the search for technology or in the application of technology. The performance of AIS processes is thus determined by the strength of the linkage of the actors and the context within which the AIS are situated, formed and maintained (Mbo'o-Tchouawou *et al.*, 2016 and Daane, 2010).

Taking holistic innovations systems perspective opens up possibilities to understand not only technological innovation, but also innovations in food systems, markets, incentives as well as local dynamics and power structures affecting them. Therefore, involving small-scale farmers in the innovation process provides opportunities for them to get solutions that are most suitable to address their problems in their local context. Hence, the use of innovation platforms (IPs) has become a popular approach to engaging smallholder farmers (Davies *et al.*, 2018) and women in particular.

Innovation platforms have increasingly become a popular development approach based on creating multi-stakeholder forums and foregrounding co-learning (Dror *et al.*, 2016; Farnworth and Colverson, 2015; Swaan *et al.*, 2013). An innovation platform is defined as a forum for learning and action involving a group of actors with different backgrounds and interests such as farmers, agricultural input suppliers, traders, food processors, researchers and government officials (Homann-Kee Tui *et al.*, 2013). These actors come together to develop a common vision and ways to achieve their goals as a group or individually (Swaans *et al.*, 2013).

Innovation Platforms (IPs) are widely viewed as a promising vehicle for increasing the impact of agricultural research and development (van Mierlo and Totin, 2014; van Paassen *et al.*, 2014). IPs are built on experiences with well-known multi-stakeholder approaches such as Farmer Field Schools, participatory research, learning alliances, local Agricultural Research Committees and Natural Resource Management Platforms (Dror *et al.*, 2016). In the field of Agricultural Research for Development (AR4D), innovation platforms form an important element of a commitment to more structural and long-term engagement among stakeholder groups (Sumberg *et al.*, 2013). Innovation platforms aim to foster agricultural innovation by facilitating and strengthening interaction and collaboration in the networks of farmers, extension officers, policy makers, researchers, Non-Governmental Organizations (NGOs), development donors, the private sector and other stakeholders (Dror *et al.*, 2016).

Following the recognition of the need for engaging diverse actors in agricultural innovation, various efforts have been made by development partners to introduce interventions which employ innovation platforms as an approach to multi-stakeholder engagement. In Tanzania, for example, the ACGG project established innovation platforms as a means to enable multi-stakeholder engagement in a bid to transform the chicken value chain.

Over the past years, innovation platforms have increasingly been established within the framework of AR4D initiatives (Dror *et al.*, 2016). Recent attempts to analyse challenges, best practices and ways to evaluate the efficiency of IPs have proved useful, therefore seeking for more insights into their effectiveness is indispensable (Cadilhon, 2013; Davies *et al.*, 2018). Documentation of and learning from the effectiveness and impact of IPs is imperative (Lundy *et al.*, 2013).

Various scholars published good case studies of IPs over the past decade such as Nederlof *et al.* (2011); Nederlof and Pyburn (2012). Most of their studies focused the innovation platforms in livestock projects, for instance crop-livestock integration in Uganda, improved natural resource management in highlands of Ethiopia, organic white gold (Cotton *spp.*) in India and milk innovation platform in Indian's Himalayan Mountains (Dror *et al.*, 2016). Others included innovation platforms such as oil palm in Ghana, cowpea and soybean in Nigeria, maize in Rwanda, maize and legumes in Nigeria, vegetables in Malawi, agriculture in Benin, oil seed in Uganda and mango in Kenya (Nederlof *et al.*, 2011).

In Tanzania, scholars have been focusing on the poultry subsector, particularly on projects aimed at developing indigenous poultry enterprises (Nederlof *et al.*, 2011). As such, little is known about innovation platforms for increasing productivity and women empowerment. In order to fill this knowledge gap, this paper aimed at assessing the strengths and challenges of community innovation platforms established by ACGG. ACGG is a research for development partnership project implemented in Tanzania, Ethiopia and Nigeria. It aims to develop public-private partnership that contributes in improving chicken productivity and promoting women empowerment. The African Chicken Genetic Gain (ACGG) project is unique in the sense that it involves multi-stakeholder collaboration by placing more emphasis on the private sector engagement. The innovation platform formed the basis for catalysing private-sector-led transformation of the chicken value chain (ACGG, 2015).

4.2   Innovation System Theory

This paper adopted Innovation System Theory. Innovation System Theory highlights the need for more communication and collaboration among heterogeneous groups and making use of their diverse capacities (knowledge, skills, capabilities, interests, resources) to address the existing challenges to innovation and take advantage of potential opportunities (Sell *et al.*, 2018; Lundvall, 2010; Malhotra *et al.*, 2009). Innovation platforms (IP) are based on innovation systems thinking. They are considered as a holistic and comprehensive framework for understanding innovation. In agriculture, IPs can be useful in exploring strategies that can boost productivity, managing natural resources sustainably and improving the value chains or influencing policies (Malhotra *et al.*, 2009).

Innovation Platforms (CIPs) comprise characteristics related to Gender Transformative Approaches (GTA). Gender transformative approaches aim at integrating gender into development, seeking to act on the social context and create an enabling environment for gender equality and women empowerment. They critically examine gender norms and inequalities; strengthen norms that support gender equality and challenge the underlying social structures, policies and norms that warrant gender inequalities. Gender Transformative Approaches (GTAs) emphasize the need for engaging a number of actors beyond women (Galiè and Kantor, 2016), which is also a characteristic of Innovation System Theory. In this paper, Community Innovation Platforms (CIP) provided an enabling environment whereby diverse actors including women collaborated in the identification of challenges and co-creation of solutions for their problems in the context of the chicken value chain and women empowerment.

4.3   IPs Conceptual Framework

The conceptual framework of this paper was adopted from African Chicken Genetic Gain (ACGG). ACGG established an innovation platform at the national level by using the analogy of organizing an African wedding. The participants of the first National Innovation Platform were engaged in analysing the most critical functions of a successful wedding. They were divided into different teams with different functions, food and beverage team, entertainment and music committee, finance committee, transport committee, guest reception committee, security committee, venue and decoration committee, video and photography committee, wedding certification committee, invitation and protocol, presents/gifts committee and logistic team. The functions to deliver the wedding are undertaken by actors whose selection is based on certain criteria. Usually, a steering committee decides who performs the aforementioned functions, typically based on a combination of two principles, namely volunteerism or provision of resources and past experience. This collaboration/team work is concerned with 'What is in it for me' (WIII-FM) consideration, which is the binding factor that keeps them interested. It is often not done because people love each other, but because they benefits from the activity, e.g. they being all alone in their own events, they draw financial rewards from it, the give back some time for social good (ACGG, 2015). Therefore, the analogy of an African wedding provides a simple framework on how actors at various levels can perform various roles towards achievement of a common goal.

A National Innovation Platform was convened on July 2015 with the aim of introducing Innovation Platform (IP) concepts and processes, identifying challenges in the chicken value chain, and the functions of the actors in the value chain. The innovation platform was based on the IP conceptual framework showing the innovation challenges, sub-challenges, value chain functions and potential actors as presented in Figure 4.1.

The Value Chain (VC) innovation challenges proposed in the first meeting were: how to multiply and avail quality chicken strains at a price smallholder farmer can afford; how to identify, continuously multiply and avail appropriate chicken strains to smallholders and establish a functional VC; How to establish a functional chicken VC that provides for the specific needs of smallholder chicken producers and enhances women participation.

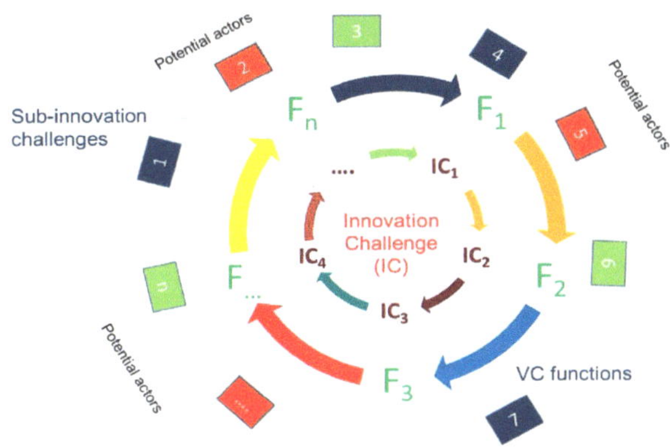

Figure 4.1:   *IP Conceptual framework*

Source: (ACGG, 2015)

In this study, the innovation platform framework established at the national level was used to establish community innovation platform which aimed at looking practical solutions related to challenges facing actors along the chicken value chain in Bariadi and Muheza Districts.

4.4   Community Innovation Platforms (CIPs) in Bariadi and Muheza Districts

The results in Table 3.1 indicate the status of CIPs in six villages namely Kwaisaka, Nkumba Kisiwani, Mlingano, Mwamoto, Byuna and Ibulyu. The number of CIPs that were conducted per year ranged from 2 to 5. One of the key informants reported that:

> Each of the three villages of Kwaisaka, Nkumba Kisiwani and Mlingano arranged their own community innovation platforms. Each village identified its own innovation challenges for instance at Kwaisaka: seasonal eruption of chicken diseases, Kisiwani Nkumba: lack of permanent poultry market and lack of poultry clinic, Mlingano: inaccessibility of extension services (ACGG Eastern zonal coordinator, 09.05.2018).

Another key informant added:

> The common innovation challenges to the three villages were marketing of chicken products (eggs and meat), non-availability of veterinary services and access to initial stocks of chicks" (ACGG Northern zonal coordinator, 05.06.2018).

These excerpts imply that innovation challenges in all villages were within the chicken value chain. Each village identified the most common innovation challenges. However, through innovation platforms, various innovation challenges along the chicken value chain could be addressed to meet the diverse interests of the actors.

Table 4.1: **Status of CIPs in Muheza and Bariadi districts**

Districts	Villages	No. of CIPs	Number of CIPs participants		Actors
			Men	**Women**	
MUHEZA	Kwaisaka	4	23	33	Men, Women, Extension Officers, Veterinary Officers, village leaders, ACGG zonal coordinators, CDOs, LFOs, brooder and Councillors
	Nkumba Kisiwani	5	17	40	Men, Women, Extension Officers, Veterinary Officers, village leaders, ACGG zonal coordinators, CDOs, LFOs, brooder and Councillors
	Mlingano	3	12	35	Men, Women, Extension Officers, Veterinary Officers, village leaders, ACGG zonal coordinators, CDOs, LFOs, brooder and Councillors
BARIADI	Ibulyu	3	16	30	Men, Women, Extension Officers, Veterinary Officers, village leaders, ACGG zonal coordinators, CDOs, LFOs and Councillors
	Mwamoto	3	16	26	Men, Women, Extension Officers, Veterinary Officers, village leaders, ACGG zonal coordinators, CDOs, LFOs and Councillors
	Byuna	2	18	29	Men, Women, Extension Officers, Veterinary Officers, village leaders, ACGG zonal coordinators, CDOs, LFOs and Councillors

The findings revealed that the extension officers under the Local Government Authority were serving as facilitators of the community innovation platforms in Bariadi and Muheza districts (LGA). Similar findings are reported by Swaans *et al.* (2013) whose study found that extension agents may be selected to play a facilitation role of the innovation platforms. Serving as facilitators of the innovation platforms provided the extension offices an opportunity to understand the capacities and interests of their clients and find the best way to engage them in collective actions in addressing their challenges.

4.5   Strengths of the community innovation platforms

4.5.1 Women participation in community innovation platforms

Community Innovation platforms were introduced to promote engagement of various actors in the chicken value chain. More emphasis was placed on women participation in the community innovation platforms. One of the key informants reported that:

> the rule that 'everyone should speak' was set in order to ensure women participate in community innovation platforms (Extension officer, Muheza District, 06.05.2018)

The findings are thus inconsistent with Farnworth and Colverson (2015) who showed that women are often underrepresented in innovation processes in terms of absolute participation as well as effective voice. The results imply that, although women were given the opportunity to speak in the innovation platforms, they merely spoke about the socio-cultural settings that influence participation in the innovation platforms and at the community level. For women to effectively participate in the innovation platforms, a specific framework to enable various actors to unpack issues associated with gender relations that contribute to women's subordinate position in the community is required.

Interestingly, the ACGG project designed a gender strategy which provided a framework for unpacking gender issues at different levels. The ACGG gender strategy comprised four dimensions of gender relations at the household level and other levels of the chicken value chain. These dimensions included: gender division of labour in chicken production; access to resources (women's and men's constraints in chicken production); control over benefits, with focus on intra household decision making and gender norms and values (Newton and Danielsen, 2017). However, the gender strategy was constrained by limited knowledge of the stakeholders on gender issues. Therefore, the facilitators of the Community Innovation Platforms require training on the gender issues, specifically on how to use the gender strategy in facilitating innovation.

4.5.2 Opportunity for multi-stakeholder's collaboration

The innovation platforms provided an opportunity for multi-stakeholder collaboration to increase chicken productivity and promote women empowerment. Among others, women were encouraged to effectively participate in community innovation platforms. Similar results have been reported by Galiè and Kantor (2016) who found that gender inclusiveness provides a potential opportunity for women empowerment. This implies that challenging the social settings that hinder women empowerment requires collective efforts of diverse stakeholders, including women themselves. Engagement of different stakeholders through the innovation platforms offers a chance for them to understand the gender issues compounding women empowerment. Moreover, innovation platforms serve as the basis for co-creation of solutions to the common challenges in the community.

4.6 Challenges of the Community Innovation Platforms

4.6.1 Limited facilitation skills

The study findings showed that the role of facilitating community innovation platforms was assigned to the village extension officers. The same has been reported by Swaans *et al* (2013) whose findings showed that agricultural extension workers can play well the facilitation role as they are more conversant/ familiar with existing agricultural systems. However, the village extension officers had limited knowledge and skills on how to facilitate community innovation platforms. This was affirmed by the Village Extension Officer from Ibulyu village who said:

> ...regardless of the one-week training offered by the ACGG project on how to facilitate innovation platforms, I am not competent enough to facilitate community innovation platforms (Village extension officer, Ibulyu village, 24.05.2018).

The results thus imply that inadequate facilitation skills of the extension officer could affect the engagement of various stakeholders in innovation platforms. Stakeholders in the innovation platforms have divergent, competing and conflicting interests. Therefore, they need skilful facilitators who can take into account their varied diverse interests. According to Tenywa *et al.* (2013) and Baruchara *et al.* (2013), facilitation is a master key in driving multi-stakeholder processes. Therefore, a continuous fostering in terms of facilitation is required to ensure successful functioning of IPs.

4.6.2 Inadequate motives of actors towards Innovation Platforms

According to the extension officer of the Kwaisaka village, the participants of the innovation platforms were mainly ACGG beneficiaries at the beginning. Later on, other members from the community joined the meetings as they realised that the platforms provided space for co-learning and co-creation of solutions to their challenges. For example, one of the IP members in Muheza District said:

> IPs helped me to access information regarding the service provider known as Silverlands who provides chicks (Non-ACGG beneficiary, 18.05.2018).

According to the extension officers, the IP members who benefited from Silverlands were not among the beneficiaries of ACGG. The results imply that interaction among the actors in the innovation platforms gave them access to information that could be useful in addressing their challenges.

Moreover, the study findings revealed that stakeholders who had not yet realised the benefits of the innovation platforms were reluctant to join those platforms. For example, one of the key informants in Muheza District said:

> I don't have time to attend village meetings, because selling of agro-products requires me to be available most of the times (Agro dealer, Muheza, 09.05.2018).

Another key informant in Bariadi District reported that:

> I am not clear with the benefits of attending community innovation platforms (Agro dealer, Bariadi, 27.05.2018).

The above excerpts imply that the actors would take part in community innovation platforms only if they expect some direct or indirect benefits from it. The same has been reported by Nederlof *et al.* (2011) whose findings showed that input dealers, traders and processors engage in the innovation platforms only if their interests are taken into consideration. However, taking into account the diverse interests in the innovation platforms is not an easy task because actors have different experiences, knowledge and skills on how to determine some expected benefits. Therefore, the facilitators should be knowledgeable on how to explore the interests of various stakeholders so as to motivate them to join innovation platforms. For example, the extension officer of Muheza District reported that inaccessibility of livestock medications and vaccines was one of the common challenges faced by the majority of livestock keepers, especially for those residing in remote areas far from the main roads. Therefore, agro-dealers can arrange their own market and ensure periodical supply of veterinary drugs, vaccines and other inputs to the communities.

4.6.3 Innovation Platforms (IPs) dependence on project funds

The responses from key informants revealed that organization and implementation of the community innovation platforms was supported by the ACGG project. Schut *et al.* (2016) report a similar experience depicting that success of innovation platforms requires provision of logistical support, backstopping of events and ensuring work accountability. Nederlof *et al.* (2011) point out that most of Innovation Platforms are funded by donors but this kind of funding will eventually end. The organizational support for innovation platforms mostly depended on external funds. However, the recognition of the potential benefits of innovation platforms calls for local actors like Local Government Authorities (LGAs) to invest their resources in strengthening the innovation platforms. Mobilization of local resources (facilitators and materials) provides an alternative access to resources for the organization and implementation of innovation platforms. Local contributions can create a sense of ownership of the innovation platforms and hence promote sustainability.

4.7   Conclusions and Recommendations

4.7.1 Conclusion

Community Innovation Platforms (CIPs) are characterised by the use of gender transformative approaches. Therefore, CIPs provide opportunities for transformation of the social context and creation an enabling environment for gender equality and women empowerment. Community Innovation Platforms (CIPs) played a significant role, especially in giving women an opportunity to meet with diverse stakeholders, discuss various challenges in the chicken value chain and co-create solutions to their challenges.

Despite the potential role of the innovation platforms in the context of improved chicken interventions, they are a new approach which requires stakeholders' understanding of the basic concepts and its applicability in rural settings.

The effectiveness of CIPs depends on adequacy of the facilitation skills for engaging multi-stakeholders in addressing challenges constraining improved chicken productivity and power relations among women and men.

4.7.2 Recommendations

Multi-stakeholder engagement in the community innovation platforms is an essential strategy for reducing poverty and promoting women empowerment particularly by transforming social structures such as beliefs and norms which impede women empowerment. Given the importance of IPs, the following recommendations have been suggested:-

Local actors such as LGAs, private sectors and other development partners should invest enough resources in establishing and strengthening community innovation platforms; Local Government Authorities (LGAs) should offer regular training on facilitation skills of innovation platforms. Such training is to be given, particularly to the Community Development Officers, Extension officers, livestock officers, village leaders and influential people.

Local Government Authorities (LGAs) and other development partners should use IPs as a means to transform the Value Chain (VC). This will help to make sure the diverse interests of the stakeholders particularly women across the value chain nodes are considered.

REFERENCES

ACGG, African Chicken Genetic Gain (2015). *Workshop on how better integrate smallholder chicken value chain in Tanzania.* 1ST National Innovation Platform. Dar es Salaam. 72pp.

Agrest, A. and Finlay, B. (2009). *Statistical Methods for the Social Sciences (Fourth Edition).*Peason Prentice Hall, New Jersey. 609pp.

Akite, I., Aryemo, I. P., Kule, E. K., Mugonola, B., Kugonza, D. R. and Okot, M. W. (2018). Gender dimensions in the local chicken value chain in northern Uganda. *African Journal of Science, Technology, Innovation and Development* 1(1): 1-14.

Akter, S., Rutsaert, P., Luis, J., Me Htwe, N., Su San, S., Raharjo, B. and Pustika, A. (2017). Women's empowerment and gender equity in agriculture: A different perspective from Southeast Asia. *Food Policy* 69: 270-279.

Alexander, A. and Welzel, C. (2007). Empowering Women: Four theories tested on four different aspects of Gender Equality. *Paper presented at the annual meeting of the Midwest Political Science Association.* Palmer House Hotel, Chicago. 39pp.

Aslam, M. (2013). Empowering women: Education and the pathways of change. *Education for All /Global Monitoring Report.* Teaching and learning: Achieving quality for all. 69pp.

Assaad, R., Nazier, H. and Ramadan, R. (2014). Individual and Households determinants of women empowerment: Application to the case of Egypt. *Working Paper 867. The Economic Research Forum (ERF).* Egypt. 36pp.

Australian Government (2016). *Gender Equality and Women's Empowerment Strategy.* Department of Foreign Affairs and Trade. Australia. 45pp.

Babbie, E. (1990). *Survey Research Methods.* Second Edition Wadworth Publishing Co. Belmont California. 395pp.

Bailey, K. D. (1994). *Methods of Social Research.* (4th Ed.) New York: The Free Press. 345pp.

Barbour, R. (Ed.) (2011). *Doing Focus Groups*. Sage Publications Ltd, Los Angelos, London, New Delhi, Singapore, and Washington DC. 174pp.

Bashir, M. K. and Schilizzi, S. (2013). Determinants of rural household food security: a comparative analysis of African and Asian studies. *Journal of the Science of Food and Agriculture* 93(6): 1251-1258.

Bryman, A. (2004). *Social Research Methods* (Second Edition). Oxford University Press, Oxford. 592pp.

Buruchara, R., Tenywa, M. M., Mugabo, J. R., Chiuri, W., Fatunbi, A. O., Adewale, A. A., Nyamwaro, S. O. and Majaliwa, M. (2013). Establishment and Implementation of Integrated Agricultural Research for Development in Eastern and Central Africa: Some Operations and Lessons Learnt from the Lake Kivu Pilot Learning Site. In; Adekunle, A. A., Fatunbi, A. O., Buruchara, R. and Nyamwaro, S. eds. Integrated Agricultural Research for Development: from Concept to Practice. Forum for Agricultural Research in Africa (FARA). pp. 120-133.

Chaudhary, A. R., Chani, M. I. and Pervaiz, Z. (2012). An analysis of different approaches to women empowerment: A case study of Pakistan. *World Applied Sciences Journal* 16(7): 971-980.

Cochran, W. G. (1977). *Sampling Techniques* (3rd ed). New York: John Wiley & Sons. 86-96pp.

Cornwall, A. (2016). Women's empowerment: What works? *Journal of International Development* 28: 342-359.

Creswell, J. W. (2013). *Research Design Qualitative and Mixed Methods Approaches 4th ed.* SAGE Publications Inc., United State of America 342pp.

Daane, J. (2010). Enhancing performance of agricultural innovation systems. *Rural Development News* 1: 76-82.

Davies, J., Maru, Y., Hall, A., Abdourhamane, I. K., Adegbidi, A., Carberry, P., Dorai, K., Ennin, S. A., Etwire, P. M., McMillan, L., Njoya, A., Ouedraogo, S., Traoré, A., Traoré-Gué, N. J. and Watson, I. (2018). Understanding innovation platform effectiveness through experiences from west and central Africa. *Agricultural Systems* 165: 321-334.

de Bruyn, J. (2018). Healthy chickens, healthy children? Exploring contributions of village poultry–keeping to the diets and growth of young children in rural Tanzania. A Thesis for Award Degree of Doctor of Philosophy of University of Sydney, Australia. 221pp.

Diiro, G. M., Seymour, G., Kassie, M., Muricho, G. and Murithi, B. W. (2018). Women's empowerment in agriculture and agricultural productivity: Evidence from rural maize farmer households in Western Kenya. *PLoS ONE* 13(5).

Dror, I., Cadilhon, J., Schut, M., Misiko, M. and Maheshwari, S. (2016). Innovation platforms for agricultural development. Evaluating the mature innovation platforms landscape. 190pp.

FAO, Food and Agriculture Organization of the United Nation (2011). The State of Food and Agriculture (SOFA) 2010-2011-Women in agriculture: Closing the gender gap for development. Rome Italy. 160pp.

FAO, Food and Agriculture Organization of the United Nation (2014). Family poultry development - Issues, opportunities and constraints. Animal production and health working paper no. 12. Rome, Italy. 33pp.

Farnworth, C. R. and Colverson, K. E. (2015). Building a Gender- Transformative Extension and Advisory Facilitation System in Sub-Saharan Africa. *Journal of Gender, Agriculture and Food Security* 1(1): 20-39.

Fox, L., Wiggins, S. Ludi, E. and Mdee, A. (2018). The lives of rural women and girls. What does an inclusive agricultural transformations that empowers women look like? ODI. 157pp. [https://news.itu.int › spotlight-digital-inclusion-girls-women-rural-thailand] site visited on 5/08/2018.

Galiè, A. and Kantor, P. (2016). From gender analysis to transforming gender norms: Using empowerment pathways to enhance gender equity and food security in Tanzania. IN: Njuki, J., Parkins, J. and Kaler, A. (eds). Transforming Gender and Food Security in the Global South. London: Routledge. 326pp.

Gupta, K. (2017). Globalization and Women Empowerment. *Journal of Social Sciences and Multidisciplinary Management Studies* 1(1): 1-4.

Haghighat, E. (2014). Establishing the connection between demographic and economic factors and gender status in the Middle East: Debunking the perception of

Islam's undue influence. *International Journal of Sociology and Social Policy* 34(7/8): 455-484.

Heshmati, A. (2017). Studies on economic development and growth in selected African countries: [http://dx.doi.org/10.1007/978-981-10-4451-9] site visited in 16th January, 2019.

Homann-Kee Tui, S., Adekunle, A., Lundy, A., Tucker, J., Birachi, E., Klerkx, L., Ballantyne, D. A., Cadilhon, J. and Mundy, P. (2013). 'What are IPs?' Brief 1 of IP practices briefs by CGIAR research Program on the Humid Tropics. ILRI. Kenya. 190pp.

Hossain, M. and Jaim, W. M. H. (2011). *Empowering Women to Become Farmer Entrepreneur*: Case Study of a NGO Supported Program in Bangladesh. Paper presented at the IFAD Conference on New Directions for Smallholder Agriculture 24-25 January, 2011, Rome, Italy. 42pp.

Huis M. A., Hansen N., Otten, S. and Lensink, R. (2017). A three-dimensional model of women's Empowerment: Implications in the field of microfinance and future directions. *Front. Psycho* 8(1678): 1-14.

Hunt, A. and Samman, E. (2016). *Women's Economic Empowerment*. Navigating enabler and constraints. London. 39pp.

Islam, N., Ahmed, E., Chew, J. and D'Netto, B. (2012). Determinants of empowerment of rural women in Bangladesh. *World Journal of Management* 4(2): 36-56.

James, V.U. (2018). *Capacity Building for Sustainable Development*. Wallingford, Oxfordshire; Boston, MA: CABI. 298pp.

Jeckoniah, J., Nombo, C. and Mdoe, N. (2012). Determinants of Women Empowerment in Agricultural in Onion Value Chains: A Case of Simanjiro District in Tanzania. *Journal of Economic and Sustainable Development* 3(10): 89-99.

Johnson, N. L., Balagamwala, M., Pinkstaff, C., Theis, S., Meinzen-Dick, R. S. and Quisumbing, A. R. (2017). How do agricultural development projects aim to empower women? Insight from an analysis of project strategies (Vol. 01609). International Food Policy Research Institute. 36pp.

Kabeer, N. (1999). Resources, agency, achievements: Reflections on the measurement of women's empowerment, *Development and Change* 30 (1): 435-464.

Kabeer, N. (2005). Gender equality and women's empowerment: A critical analysis of the third millennium development goal 1, *Gender and Development* 13(1): 13-24.

Kishor, S. and Gupta, K. (2004). Women's empowerment in India and its states. Evidence from the NFHS. *Economic and Political Weekly* 39(7): 694-712.

Kumar, C. A. (2015). Women empowerment and its impacts on efficiency of women executives and organization. Thesis for Award Degree of PhD at Acharya Nagarjuna University, India. 264pp.

Lee, M. and Finlay, J. (2017). The Effect of Reproductive Health Improvements on Women's Economic Empowerment: A Review through the Population and Poverty (POPPOV) Lens (Washington, DC: Population and Poverty Research Initiative and Population Reference Bureau). 24pp.

Leroy, G., Baumung, R., Boettcher, P., Scherf, B. and Hoffmann, I. (2015). Review: Sustainability of crossbreeding in developing countries; definitely not like crossing a meadow. FAO, Rome, Italy. 12pp.

Lundvall, B. A. (2010). *National Systems of Innovation: Towards A Theory of Innovation and Interactive Learning* (New edition). London. 388pp.

Malamsha, K. C. T. (2014). Success of Microfinance Institutions in Tanzania the Case of Savings and Credits Co-operative Societies. A Thesis for Award Degree of Doctor of Philosophy of Sokoine University of Agriculture, Morogoro. 223pp.

Malamsha, K. C. T. (2014). Success of Microfinance Institutions in Tanzania the Case of Savings and Credits Co-operative Societies. A Thesis for Award Degree of Doctor of Philosophy of Sokoine University of Agriculture, Morogoro. 223pp.

Malapit, H., Quisumbing, A., Meinzen-Dick, R., Seymour, G., Martinez, E. M., Heckert, J., Rubin, D., Vaz, A. and Yount, K. M. (2019). Development of the project-level Women's Empowerment in Agriculture Index (pro-WEAI). IFPRI Discussion paper 1796. Washington, DC: International Food Policy Research Institute. 57pp.

Malhotra, A., Schuler, R. S. and Boender, C. (2002). Measuring women empowerment as a variable in international development. Background paper prepared for the world bank workshop on poverty and gender: New Perspectives. Final version. 58pp.

Malhotra, A., Schuler, R. S. and Boender, C. (2002). Measuring women empowerment as a variable in international development. *Background paper prepared for the world bank workshop on poverty and gender: New Perspectives*. Final version. 58pp.

Malhotra, A., Schulte, J., Patel, P. and Petesch, P. (2009). *Innovation for Women's Empowerment and Gender Equality*. International Center for Research on women.14pp.

Mangino, K. M. B. (2016). Who defines women's empowerment: Exploring global, national and local definitions through an Indonesian case study. *Journal of the Graduate School of Asia-Pacific Studies* 32: 113-129.

Mathialagan, P. (2015). Women empowerment through participation in backyard poultry Rearing: A case Study from Namakkal District, Tamil Nadu. *The Indian Veterinary Journal* 92(6): 75-76.

Mathialagan, P. (2014). Integration of innovations in backyard poultry rearing for empowering rural women. *International Journal of Science, Environment and Technology* 3(6): 2190-2196.

Mbo'o-Tchouawou, M., Waithanji, E., Mulei, L. and Karugia, J. (2016). Using an analytical model to explore potential gendered dimensions in agricultural innovation systems. ReSAKSS Working Paper No. 39. 2016. 28pp.

Meinzen-Dick, R., Rubin, D., Elias, M., Mulema, A. A. and Myers, E. (2019). *Women's Empowerment in Agriculture: Lessons from Qualitative Research*. IFPRI. Washington DC. 56pp.

Nederlf, E. S. and Pyburn, R. (eds) (2012). One Finger Cannot Lift a Rock: Facilitating Innovation Platforms to Trigger Institutional Change in West-Africa. KIT-Royal Tropical Institute, Amsterdam, the Netherlands. [https://www.kit.nl › staff › gerard-baltissen] site visited on 12/08/2018.

Nederlof, S., Mariana, W. and Femke, V. (2011). Putting heads together. Agricultural innovation in practice. Bulletin 396, KIT Publisher. Development, Policy & Practice. 14pp.

Newbury, E. (2017). Understanding women's lives in polygamous marriages: Exploring community perspectives in Sierra Leone and DRC. Trocaire, Ireland. 24pp.

Newton, J. and Danielsen, K. (2017). Gender strategy for African Chicken Genetic Gain program (ACGG). KIT/Sustainable Economic Department and Gender. 29pp.

Njoh, A. J. and Akiwumi, F. A. (2012). The impact of religion on women empowerment as a millennium development goal in Africa. *Social Indicator Research* 107: 1-18.

Nyange, T. M., Sikira, A. N. and Lyimo-Macha, J. G. (2017). Legal aid service interventions on women Empowerment in Morogoro rural and Kongwa districts, Tanzania. *International Journal of Asian Social Science* 7(7): 570-583.

Okitoi, L. O., Ondwasy, H. O., Obali, M. P. and Murekefu, F. (2007). Gender issues in poultry production in rural households of Western Kenya. *Livestock Research for Rural Development. Volume 19, Article #17.* Retrieved, from [http://www.lrrd.org/lrrd19/2/okit19017.htm] site visited on October 25, 2018.

Pan, S. (2017). Women Empowerment – A strategy for Development. *International Research Journal of Multidisciplinary Studies* 3(7): 1- 5.

Peterman, A., Schwab, B., Roy, S., Hidrobo, M. and Giligan, D. (2015). Measuring Women's Decision making: Indicator and survey design experiment from cash and food transfer evaluations in Ecuador, Uganda and Yemen, "IFPRI Discussion paper 1453. 52pp.

Quisumbing, A. R., Roy, S., Njuki, J., Tanvin, K. and Waithanji, E. (2013). Can daily value-chain projects change gender norms in Rural Bangladesh? Impacts on assets, gender norms and time use. *Discussion Paper 1311.* Washington, DC: International Food Policy Research Institute. 4pp.

Santos, F., Fletschner, D., Savath, V. and Peterman, A. (2014). Can government-allocated land contribute to food security? Intrahousehold analysis of West Bengal's microplot allocation program. *World Development* 64: 860-872.

Schut, M., Cadilhon, J., Misiko, M and Dror, I. (2016). The state of innovation platforms in agricultural research for development. Routledge, New York. 15pp.

Sell, M., Vihinen, H., Gabiso, G. and Lindström, K. (2018). Innovation platforms: a tool to enhance small scale farmer potential through co-creation, *Development in Practice* 1: 1-13.

Sheikh, Q. A., Meraj, M. and Sadaqat, M. (2016). Gender equality and socio-economic development through women empowerment in Pakistan. *Journal of Asia Pacific Studies* 34: 142-160.

Sikira, A. N. (2010). Women empowerment and gender based violence in Serengeti district, Mara region, Tanzania. Thesis for Award Degree of PhD at Sokoine University of Agriculture, Morogoro, Tanzania. 207pp.

Swaans, K., Cullen, B., van Rooyen, A., Adekunle, A., Ngwenya, H., Lemma, Z. and Nederlof, S. (2013). Dealing with Critical Challenges in Innovation Platforms: Lessons for Facilitation, *Knowledge management for Development Journal* 9(3): 116-135.

Taylor, G. and Pereznieto, P. (2014). *Review of Evaluation Approaches and Methods Used by Interventions on Women and Girls' Economic Empowerment.* London: ODI. 49pp.

Tenywa, M. M., Farrow, A. Buruchara, R., Tukahirwa, J. M. B., Mugabe, J., Rao, K. P. C. Wanjiku, C., Nyamwaro, S. O., Kashaija, N. I, Majaliwa, M., Mapatano, S., Mugabo, J., Ngaboyisonga, C. M., Ramazine, M. A., Mutabazi, S., Fungo, B., Pali, P., Njuki, J., Abenakyo, A., Opondo, C., Nkonya, E., Njeru, R., Lubanga, L., Wimba, B, Murorunkwere, F., Kuule, M., Mandefu, P., Kamugisha, R., Fatunbi, A. O. and Adekunle, A. A. (2013). Strategies for Setting up Innovation Platforms in the Lake Kivu Pilot Learning Site. In Adekunle, A. A., Fatunbi, A. O., Buruchara, R., Nyamwaro, S. eds. *Integrated Agricultural Research for Development: from Concept to Practice.* Forum for Agricultural Research in Africa (FARA). Accra, Ghana. pp 20-41.

Tesfaye, B., Mulema, A., Kinati, W. and Lemma, M. (2018). Stakeholders consultation on gender and livestock. *Ethiopia partners workshop report.* CGIAR. Ethiopia. 18pp.

Thieme, O., Sonaiya, F., Rota, A., Gueye, F., Dolberg, F. and Alders, R. (2014). Defining family poultry production systems and their contribution to livelihoods. In: *Decision Tools for Family Poultry Development*. FAO Animal Production and Health Guidelines No. 16. Rome, Italy. pp 3-8.

UN, United Nations (2014). *Beijing Declaration and Platform for Action*. Beijing +5 political declaration and outcome. UN women. New York. 270pp.

UN, United Nations (2015). Commission on the status of women. Report on the fifty-ninth session (21 March 2014 and 9-20 March 2015). New York. 46pp.

UN, United Nations (2015). Transforming our world: The 2030 agenda for sustainable development. 41pp.

UNDP, United Nations Development Programme (2014). Human Development Report: Sustaining Human Progress Reducing Vulnerability and building Resilience. New York. 239pp.

URT, United Republic of Tanzania (2012). *National Sample Census of Agriculture 2007/2008*. Small holder Agriculture. Volume iii Livestock Sector National Report. National Bureau of Statistics. 176pp.

van Eerdewijk, A., Wong, F., Vaast, C. Newton, J., Tyszler, M. and Pennington, A. (2017). White Paper: A conceptual model of women and girls' empowerment. Royal Tropical Institute (KIT) Amsterdam. 82pp.

van Mierlo, B., Totin, E. (2014). Between script and improvisation: Institutional conditions and their local operation. *Outlook on Agriculture* 43: 157-163.

van Paassen, A. Klerkx, L., Adu-Acheampong, R., Adjei-Nsiah, S. and Zannoue, E. (2014). Agricultural innovation platforms in West Africa: How does strategic institutional entrepreneurship unfold in different value chain context? *Outlook on Agriculture* 43: 193-200.

Wong, J. T., de Bruyn, J., Bagnol, B., Grieve, H., Li, M., Pym, R. and Alders, R. G. (2017). Small-scale poultry and food security in resource-poor settings: A review. *Global Food Security* 15: 43-52.

World Bank (2015). *World Bank Group Gender Strategy (FY16-23: Gender Equality, Poverty Reduction and Inclusive Growth*. Washington, DC. 97pp.

www.ingramcontent.com/pod-product-compliance
Lightning Source LLC
Chambersburg PA
CBHW040954110726
48007CB00001B/12